Beekeeping Secrets

The Safe Way to Raise Bees

By: Alicia Moore

TABLE OF CONTENTS

Alicia Moore

PUBLISHERS NOTES

Speedy Publishing LLC

40 E. Main St., #1156

Newark, DE 19711

www.speedypublishing.co

Cover Artwork: 24 Hr. Designs Ltd.

Editing: Speedy Publishing LLC

Book design: Speedy Publishing LLC

ISBN:

This is a reprint book.

DISCLAIMER

This publication is intended to provide helpful and informative material. It is not intended to diagnose, treat, cure, or prevent any health problem or condition, nor is intended to replace the advice of a physician. No action should be taken solely on the contents of this book. Always consult your physician or qualified health-care professional on any matters regarding your health and before adopting any suggestions in this book or drawing inferences from it.

The author and publisher specifically disclaim all responsibility for any liability, loss or risk, personal or otherwise, which is incurred as a consequence, directly or indirectly, from the use or application of any contents of this book.

Any and all product names referenced within this book are the trademarks of their respective owners. None of these owners have sponsored, authorized, endorsed, or approved this book.

Always read all information provided by the manufacturers' product labels before using their products. The author and publisher are not responsible for claims made by manufacturers.

Alicia Moore

DEDICATION

This book is dedicated to my immediate family especially my Dad Timothy.

CHAPTER 1- GETTING STARTED IN BEEKEEPING

If you are considering bees as a hobby or as a sideline business, there are things you will want to keep in mind before making that decision. Since there are many factors involved with making money with the honeybees produce, you might want to start doing it as a hobby. There is a significant amount of money in the start-up of beekeeping.

Before investing any amount of money in your beekeeping project, you might want contact beekeepers in your area. As a rule, they will more than happy to share their experience with you. Most beekeepers love keeping bees and to them it is just a "hobby", but they can give you some insight into beekeeping. Take plenty of notes. More likely than not you will need them.

In making the decision of becoming a beekeeper, you will want to consider the safety of family, friends, and neighbors. You wouldn't want someone to get stung that is allergic to bee stings. The best course of action on that account is to ask your neighbors and friends, if any of them are allergic to bees. You will also be able to find out if there might be someone who would not like beehives so close to their proximity. You will also want to check with the county you live in. You will want to know about any ordinances or laws prohibiting beekeeping.

You will also want to consider whether or not you have a location that would be conducive to maintaining bees. You will also want to consider where the bees will have to fly to retrieve nectar and pollen. Keeping plants they like close by is not a bad idea either. Since bees need water every day, you might want to have water for them close at hand. You don't want them visiting the neighbor's swimming pool. Here is a list of spots unacceptable to the health of the bees.

How many months of the year will pollen and nectar will be readily available to the bees?

Will you have to feed them in order for them to survive and how much of the year?

Is there a water supply available year round for the bees? They need water every day.

You will need to consider what will be underneath the bees as they fly to get the nectar and pollen they require. The bees will defecate as they are flying and their feces will leave spots on everything below them. The feces can even ruin the surface of a vehicle. There are methods to use to force the bees to fly at a higher altitude, such as a tall fence or thick tall plants near the hive.

You want the hives accessible year round.

You will want to avoid low spots for your hives because they hold the cold, damp air too long.

You will also want to avoid high spots for your hives because that would be too windy.

These are just some of the things you will want to consider before taking on this hobby.

During a nectar flow, many of the older workers will be in the field hunting for food. This is the best time to examine the colony. During the summer more bees will be in the hive and the situation can change, especially between the nectar flows. There can be some robbing going on at this time, which will make the bees even more defensive at any intrusion to their hive. Leaving the colony open for more than a few minutes can accelerate a robbing as can leaving cappings or honey exposed. It will become a necessity to reduce the entrance of a weak colony to prevent stronger hives attempt to rob from it. A honey flow will reduce the likelihood of robbing.

The mood of the bees can have a lot to do with the weather or the time of day. On the days of rainy weather, cool temperatures, early in the morning or late in the afternoon will be more likely to make them angry and they will attack. Always inspect them on warm, sunny days in the middle of the day when most of the bees are foraging.

Keep a constant warm water supply for the bees to cool the hive and dilute honey to feed t heir young. They will collect water from the closest water source. If you do not have a constant supply of shallow water for the bees, they will look for it somewhere else, like the neighbor's pool, birdbath or wading ponds. The bees are more likely to drown in those sources. If you have a water supply for them when they first fly out in spring, they will not go anywhere else for water. Once they find a water source, it is hard to keep them from going back to it.

A beekeeper must keep the bees in control every time the hive is open. A typical hive can house thousands of workers all capable of stinging. There are measures a beekeeper can take in the open that he cannot take in the city because of the closeness of other people.

Smoke is the most important tool for the beekeeper opening a hive. Smoke should be used in moderation, but the smoker should be capable of producing large volumes of smoke on short notice. The beekeeper must smoke the entrance of the hive, under the cover, and periodically smoke the frames while the hive is open.

Alicia Moore

Try not to jar the hive or the frames as that may anger the bees, which will make it hard for a beekeeper to do his work. The beekeeper must work quickly and carefully. By going through the frames several times a year, the beekeeper keeps the frames movable. Remove any excess combs.

Using gloves when working with bees make the beekeeper clumsier and he may lose control of the hive. The stings that the gloves are protecting you from are easily removed and the pain quickly passes.

CHAPTER 2- CLOTHING AND EQUIPMENT NEEDED

One of the most important pieces of clothing a beekeeper wears is the veil. Bee stings on the face can be very painful and there is the possibility of damage to the eyes and ears.

If by chance a bee gets inside the veil, walk away from the hives and remove the bees. Never remove the veil when you are in with the hives.

Use protective clothing to avoid getting hive product on your regular clothes, and to protect sensitive areas of your body. Avoid dark or rough textured clothes. Bees are able to hold on to a rough texture material than smooth material. Wear white or light colored coveralls. If you are not using boots, do not wear dark socks. Boots that fasten over the coveralls or in the coveralls should be worn. A windbreaker jacket will help you to avoid being stung. Pants, veil, sleeves should be fasten securely to prevent bees from getting into your clothes.

If a bee does get into your clothing, squeeze it in the clothing or walk away from the hives and open up your clothing to allow the bee to escape. Before handling bees, do not use any sweet smelling cologne, hair spray or any other products. The odor may irritate the bees or attract them. Glove should be used sparingly. Gloves are useful during bad weather or when moving colonies, but gloves can hinder the manipulating of the colonies. Without the interference of gloves, you will find that the bees respond better to a lighter touch.

As a beginner you will want to contemplate the number of colonies you want to start out with. Two or three is a good number to start off with because it will give you a chance to compare the two colonies, such as the growth and the production.

The equipment you will need to start off with for a complete hive is:

1 metal covered top
1 inner cover
1 bottom board
2 standard 10-framc hive bodies, each body contains 10-frames
1 queen excluder
2 shallow 10-frame supers with frames
1 bee smoker
1 hive tool
1 pr. bee gloves
1 pair coveralls
1 bee veil

You can buy this equipment new or used. If it is used you will want to make sure it is in good condition. Also have it examined by the Apiary Inspection Service for any possibility of disease. The equipment will run you $250 or more. If you are really talented and ambitious you can build your own hives. Just make sure you have the dimensions correct because bees will build combs where you least want them.

CHAPTER 3- HOW TO HANDLE BEES

Intruders are going to get stung by the bees protecting the hive. As a beekeeper you will have to be prepared to receive your share of stings. If you have any fear of bees or of being stung, you will have to conquer those apprehensions. As you gain confidence and more adept at the handling of the bees, the stings will happen less frequently.

One of the tips you will want to learn is when to manipulate bees. You can open and examine the bee colonies on days that are warm and sunny with no wind. The older bees will be out searching for food on those days. The older bees will stay in the hive on colder, windy and rainy days.

When there is an abundance of nectar, bees are much easier to examine than when there is a shortage of nectar. Plying them with sugar syrup may help, but not always.

Spring is the best time to examine the bees because of smaller populations.

Bees will usually tolerate a moderate beekeeper manipulation for 10 to 15 minutes. It is best not to keep the hives open any longer than you have to. Brood examinations should never be drawn out. When examining the hives, if bees become noisy or very nervous, the hive needs to be closed. If there is honey in the combs, they will attract robber bees unless there is an over abundance of nectar. If robbing start, you will need to stop examinations for the rest of the day and reduce the entrances to the hives. Once the robbing starts it is difficult to stop.

If you need to manipulate a colony, have a lighted smoker that omits cool smoke. Before you open the hives, you want to puff smoke into the entrance of the hive. Move on to the other colonies allowing time for the bees to react to the smoke. Keep your smoker handy because you will need it while you are making your close inspections of each colony. If you have some of the bees looking at you, make them scatter with a few puffs of smoke. When you are around the bees, you should move smoothly and carefully so that you don't alarm the bees.

When prying off the cover to the hive be as gentle as possible, bees are sensitive to vibrations. Avoid any jolting of the hives. After removing the cover to the hive, work from the back or the side of the hive. Remove the frame nearest the outside to be examined. If robbing is not a problem, lean the frame against the outside of the hive to give you more room to work. If robbing could be a problem make sure to cover the hives and never leave a frame out in the open.

If you are going to examine all the boxes, start with the lowest one. Make sure the boxes you are not examining stay covered. After examining the lowest box, examine each box after it has been replaced on the lower one.

When you need to remove the frame, pry it loose with the hive tool. With a firm grip on the loosened frame, gently lift it, trying not to scrape the bees on the adjoining frame. Leave the frame outside of the hive or box, to give you a larger working area. If you scrape the comb, do leave the bits and pieces in the hive or box. Only scrape comb that it in the way, scraping is irritating to the bees.

CHAPTER 4- ACQUIRING BEES

There are several ways to acquire bees. No matter the method you choose spring is the best time to purchase bees. Listed below are methods by which to acquiring bees.

Established Colonies

Established colonies will cost you more, but they can be worth the extra money. Before you purchase the bees have them and their equipment inspected by a state bee inspector. Dilapidated equipment or weak colonies you will want to stay away from

When purchasing established colonies, the equipment will not require any assembly. Since the queen is already laying eggs, will be able to judge her brood pattern. The chance of producing a honey crop the first year with an established colony is very good. The previous owner should be able to give you any history or background information of the bees.

If you are a beginner, a strong colony may be more than you are ready to handle. The equipment may be old and need replacing, or it may not be standard equipment.

Nucleus Colonies (Nucs)

The nucleus colony is a smaller colony of bees taken from an established colony. The "nucs" hives have fewer frames than a standard hive. The nucleus colony consists of only four or five frames instead of the standard 10-frames. They can house extra queens and they can be used to raise new queens. The nucleus colony comes with the four or five frames of brood, honey and pollen, a laying queen, and every frame should be full of adult bees.

Nucleus colonies are less expensive than established colonies. The queens are usually new, giving you the opportunity to judge her brood pattern. If the nucleus colony has a strong nectar flow, there is a possibility of a honey crop the first year. Usually they can be purchased locally. Since the nucleus colony is not as strong as an established colony, they may be easier for a beginner to handle. You still need to have them inspected for disease.

Package Bees

Package bee producers produce package bees in southern United States. The package bees consists of 2 or 3 pounds of bees, a queen in a separate cage, and a canister of sugar syrup used to feed the bees during transport. They are shipped in a special screen mailing cages through the U.S. Postal Service.

The package bees are cheaper than the established or the nucleus colonies. Beginners should be able to handle them easily. The possibility of the broods having a disease is slim.

The package bees may not produce a honey crop the first year. It will be more difficult to judge the queen with no brood. Because of the strain of being transported, a queen may be out-dated which can lead to an unproductive queen. If the weather is bad, you will have a difficult time in introducing the bees into the hives. The bees will have to be fed until the start of the nectar flow.

Swarms

Swarms can be a fun way to get bees, and they are free. They can be easily collected and placed in prepared equipment. It is usually a good idea to introduce a new queen as soon as possible to the swarm. The swarms can be rather large by they can be easily handled.

You will not get a brood so you will not be able to judge the new queen. The swarms are unlikely to produce honey crop the first year, but that does depend on the size of the swarm. The availability of swarms is very unpredictable.

Queen Management Techniques

When a colony is not performing well, it is common practice to introduce a new queen into the colony. There are certain qualities that a beekeeper looks for in a queen 's offspring, such as good collectors of honey or pollen, resistance to disease and pests, reduced swarming, gentleness, effective pollination, and minimal propolis use. Propolis is the wax-type resin derived from a tree bees use as glue.

It is a common practice to mark the queen with a small spot of paint on her back because the queen is the source of all the worker bees in the colony. They are impossible to distinguish one from another without an identifying mark. The beekeeping industry uses a color code that indicates the year the queen was introduced into the colony. Model car paint is often used to place a very small dot on the back of the queen. The queen is usually marked prior to the introduction into the colony, but she can be marked at any time. Sometime a purchased queen will come already marked. The color code used is:

White (or gray) for years ending 1 or 6
Yellow for years ending 2 or 7
Red for years ending 3 or 8
Green for years ending 4 or 9
Blue for years ending 5 or 0

The residents of the colony may reject or even kill a newly introduced queen, unless certain requirements are not met. There are several different methods that have been published over the years, but a particular procedure has not been accepted as the best procedure for all occasions. The most common practice of all the procedures requires an introductory period of about three days. The queen is placed in a cage and is fed by the colony bees though the wire gauze covering the cage. The only way she can be released is by the worker bees eating a candy entrance. The beekeeper can decide to release the queen into the colony manually.

The older more established worker bees are not as receptive as the younger bees to a new queen. You can turn the colony entrance to face the opposite direction to separate the older from the younger bees. In an empty hive place at least one frame of honey facing the original direction. The older bees will leave the original hive and return to the new empty hive. The original hive will only have the younger bees, while most of the new hive will have accumulated the older bees. The queen can then be introduced into the hive of the younger bees without problems. The two colonies can be reunited after the new queen is established.

Before introducing a new queen into a colony, make sure the colony does not have a queen, and any of the developing queen cells are destroyed. Leave the colony without the queen for a day or so. Let the queen be caged for about two days. To release a queen, place the cage between the frames with the screen side down and the candy plug exposed to the younger bees and the brood.

Allow the bees two days to release the queen and then remove the cage as soon as possible. If the queen is to be release manually, watch the surrounding bees to determine if they are clinging tightly to the cage the queen is in. If they behave in an aggressive behavior, do not release the queen until the bees act passively toward the cage. Once you have released the queen, watch closely to see if the other bees are react with hostility to the new queen as she explores the comb on which she was released. Don't open the hive again for a few days allowing the queen time to start her brood nest.

A good technique and careful handling will ensure the success of introducing a new queen into the colony. Other factors can also play a part, such as environment conditions, changing seasons, the availability of food, and beekeeper competence.

Raising Queen Bees

The success of the colony depends largely on the quality of the queen. As a beekeeper you may notice a difference in the production of honey from one colony to the next. The difference in production can depend on several factors, one of which is the queen. Beekeepers call this trait as "queenlessness". When the queen is in the state does less brood rearing, drone layers and shows queenlessness, must be replaced. When beekeepers spot this condition going on in one of his colonies he will, what is known as "requeen " the colony. Requeening is basically introducing a new queen into the colony. Although queen bees can be purchased from commercial beekeepers, but prefer to raise the queen themselves in order to continue with a queen of the strain or stock of previous queens that has produced so much success in his colonies. Purchasing queen bees from a commercial beekeeper does not guarantee a queen of from a good strain.

When rearing queens it is best to use larvae that are under 24 hours old. Larvae of this age have not been exposed to the worker's diet. It is important that the future queen larvae be fed queen jelly. Queens are raised from the same fertilized eggs as the worker bees. When the eggs are newly hatched, they are neither a queen nor a worker bee. Once the hatched larvae are 3 days old pollen is introduced into the diet of the larvae destined to become worker bees. On the other hand the hatched larvae destined to

Alicia Moore

become queen bees are raised in what is known as the queen cell which has been specially built.

There are requirements to raising a good queen. The needs to be an ample supply of nectar and good quality pollen, as well as an abundance of sexually mature, high-quality drones for mating with the newly emerge virgin queens. There must be suitable weather for mating of the drones and the queens. There needs to be a good queen mother to breed from, whose offspring worker bees (and colonies) seem to have the qualities desired, such as gentle temperament, disease resistance, low swarming tendency and excellent honey production.

This is a summary of the steps to be taken for queen raising. A starter colony must be established for the beginning of raising queen cells. A cell building colony must be established. Then there is the grafting of the honey bee larvae. Last but not lest the transferring the mature queen cells to honey bee nucleus colonies for the mating stage.

As a starter colony, choose a strong two-story colony that is headed by a two-year old queen. It will be necessary to locate and temporarily remove the queen along with the comb she is sitting on with bees, to a spare empty 8-frame box or nucleus hive. Then the 2-story hive needs to move about 2 meters to the rear of its original site.

Now you can prepare the starter colony by placing an empty box with a bottom board and the lid on the bottom of the hive. Four combs of unsealed brood with the adult bees from the two-story hive must be moved to the empty hive. Also place a comb of unsealed honey and pollen with bees on each side of the brood. Fill in the rest of the empty box with empty combs.

Take another 2 or 3 other brood combs with extra young bees and shake them into the 2-story hive. Add a feeder of sugar syrup to the starter colony. Since the bees will be what is known as "queenless", the nurse bees in the starter colony will be stimulated to feed and produce more brood food. Return the 2-year old queen and her comb to the bottom box of the 2-story hive.

The cell builder colony is another important step in raising queen bees. The aim of this procedure is to create a situation under which bees will carefully nurture the young, developing queens. You will want to select a cell builder colony that is a strong colony that fully occupies a large hive. A 3-story hive will work to your best advantage, by reducing the available space to two hives. Confine the queen to the bottom box. This brood chamber should be equipped with an equal amount of brood and empty drawn cells for the queen to lay eggs.

Two combs of very young larvae should be placed in the center of the super (the hive body) and fill in the remaining space with combs of honey and pollen. It is necessary to place the combs of unsealed honey and pollen along side of the combs of unsealed larvae. This makes it look like a natural brood nest. With the queen being confined, it will prevent her from entering into the super. Recruited nurse bees will feed the unsealed larvae in the super. The bees will soon become aware the queen is not occupying the nest. This begins the impulse of the nurse bees taking the steps to rear a new queen. This is the type of environment you will want to place newly grafted or started cells to be introduced for rearing. You will want to leave the cell building colony for 24 hours before inserting the newly grafted or started cells.

You will want to leave a space between the two brood combs in the super. The space needs to be wide enough to fit a cell bar. A cell bar is a wooden strip that holds queen cups for rearing queens.

If possible it is best not to rear queen during a heavy honey flow. A light nectar flow with ample pollen, preferably a mixture of pollens, is the best condition for rearing queens. If supplementary feeding becomes necessary, always use a mixture of 2 parts sugar to 1 part water for sugar syrup to simulate nectar. Never use diluted honey.

Grafting is the process of removing worker larvae from its cell and placing it into an artificial queen cup for rearing the larvae into a queen. You start the grafting process by preparing the bars of cells by sticking 20 plastic cups onto a wax covered board. The bar must be placed into a hive for at least 24 hours before grafting. During this time the bees will clean and condition the cell cups.

You will need a grafting tool to transfer larvae. Each larva is floating on a little raft of royal jelly and must be placed undisturbed into the bottom of the conditioned cups. The grafting tool must be able to follow the curve of the bottom of the cup to allow it to be inserted under the back of the tiny floating larva without touching it.

The best conditions to graft in is cool temperatures and well fed larvae, the priming of the cell cups with diluted royal jelly should not be necessary. Do not graft in very hot weather or in low humidity. The larvae could potential be damaged by dehydration. Only graft larvae that are under 24 hours of age from hatching and are floating on a good amount of royal jelly. Never expose the larvae to direct sunlight and work as quickly as possible.

The grafted larvae should be placed into an abundance of nurse bees that are far enough away from a queen that they will attempt rear all the cells. The age of the nurse bees range from 9 days to 12 days after they have emerged from a cell. It is always important to have a large number of replacement young bees available to the colony in order to provide nurse bees. The production of royal jelly depends on an ample supply of pollen or pollen substitutes. Lack of pollens leads to smaller, less well-fed larvae and queens. Also the nurse bees will lose their body reserves of stored nutrients and become susceptible to disease.

It is very important to record the day the cells were grafted and the day the queens are due to emerge. A queen will emerge 16 days after the egg was laid, or 13 days after the egg hatches into a larva. Since the larva was grafted at 24 hours old, the queen will emerge 12 days later. If one of the queens emerge early, she will kill all the remaining cells. It is best if the cells are left until the day before they are due to emerge, it is then possible to move the cells from the cell build colony to the nuclei.

When you are transporting the cells to the nuclei, the cells must be handled gently to avoid damage to the immature queens. Make the transition to the mating yard. Do not shake or jar the combs or bars with cells, and avoid turning the cells from the natural position. Do not allow them to be exposed to direct sunlight, and because the queen nymph is susceptible to cold do not allow the cells out of the hive too long, or exposed to cold winds or a chilly atmosphere.

Cells should be distributed to the mating yard as soon as possible after the nucleus colony has set up. You do not want too much time to lapse or the bees in the nucleus will start building cells. It will be necessary to destroy all of these cells before inserting the raised cells into the nuclei. Only one cell is given to a nucleus. A wet, sharp knife can be used to separate adjoining cells on the cell bar. Each cell must be carefully removed from the bar and placed into the nucleus hive. First a side comb is removed from the nucleus to allow room for manipulation. A small depression is pressed into the face of the center brood comb and t he plastic base of the cell gently pressed into it.

Alicia Moore

Mark every nucleus with a date the young queen is due to emerge and the mother queen she was bred from should be noted. A virgin queen will mate and start laying about 10 days after she has emerged from the cell. In the fall this period can continue longer than the normal time. Do not open or move the nucleus during the mating period. It is important that the virgin queen start mating. The mating takes place while she is flying in the open and not in the hive. The mating does not begin until the queen is sexually mature. This takes place 5 to 6 days after emerging. The queen must mate within 20 days, if not she will remain infertile. Most of the queen rears will destroy all the queens that fail to lay on time, except in the fall when mating and expected laying time can be extended because of cooler weather.

CHAPTER 5- USING NECTAR SUBSTITUTES

Using Nectar Substitutes

Plants have a glandular secretion, called nectar, which usually collects at the base of the flowers. Bees depend on this nectar for their source of energy. Honeybees dehydrate nectar to produce honey because it contains a low to moderate concentration of sugar. If a little pollen is incorporated into it, there can be barely measurable amounts of proteins, vitamins and other nutrients in the nectar.

There is two different ways bees use nectar. The nectar will work as a substitute for water, used to dilute brood food and air condition the hive. The bees can also ripen the nectar to become a stored resource for carbohydrate. The nectar substitute can also be used in either one of those ways, but the beekeeper use different sugar concentrations for different purposes.

Inspections of the colony should be conducted about every ten days during early and late spring. A beekeeper must stay aware of the conditions of the colony and the inspections will accomplish this. During the early spring the beekeeper must be aware of the food supply and if it is enough. During the late spring the beekeeper must be attentive to the possibility of swarming to keep it under control. Every inspection should inform the beekeeper if the bees have adequate food to get them through the times of bad weather. If they have enough to get them through until the next inspection, the beekeeper will again check their supply. If not, then the bees will have to be fed.

In the spring beekeepers will always feed the bees a pollen substitute and if the bees need to be fed sugar syrup. The sugar syrups fed early in the season are used for brood rearing. Feeding sugar usually stimulates egg laying and the syrup is usually a "light" syrup mixed with 1 part sugar and 1 par water. A heavy syrup, a mixture of 2 parts sugar and 1 part water, is fed late in the season to ensure adequate winter food supplies. They are stored as ripened syrup. If a medicated treatment is needed in the fall, feed for weight first, and then top off the colony with medicated syrup. There are beekeepers who use high fructose corn syrup to feed their bees, but they do not usually dilute the syrup regardless of the season. There are some levels of hydroxymethylfurfural (HMF) that will increase over time, especially with heat. HMG is toxic to honeybees at high enough concentrations.

It is best to feed the syrup to each colony individually. Every colony should receive its full share regardless of the size of the colony. It is best to feed in the evening, after the bees have settled down for the day. If there is a sudden abundance of syrup, bees will interpret this as an opportunity for robbing, by feeding after flying has ceased; the potential robbers find a source at home. Don't spill any on the hive, this will attract ants and robbing bees.

Using Pollen Substitutes

Pollen is a source of protein, vitamins, mineral and some carbohydrates for honeybees. One pollen alone does not provide a bee with all the nutrients they need to stay healthy, so a variety of pollens are needed to provide them will all the nutrients they need. Without these nutrients, bees would not be able to produce the royal jelly required to feed the queen and rear brood. If the weather will not allow the bees to leave the hive for several days to collect pollen, and there is very little stored in the combs, it will be necessary the beekeeper to feed the bees a pollen substitute. At the same time the beekeeper will feed them sugar syrup.

The main ingredient used in making a pollen substitute is brewer's yeast. The yeast can be fed to the bees dry, but the bees can better utilize the yeast when it is made into patties with the consistency of peanut butter. The yeast is often mixed with 50% sucrose syrup to moisten the patties. The patties are wrapped in wax paper or placed inside plastic bags to keep them moist. The beekeepers that use the high fructose corn syrup will mix the patties using that syrup. Other ingredients can be added to the patties that offer more nutrients than the yeast and syrup mixture alone. Beekeepers will add casein, lactalbumin or soy flour to their mixtures.

If the beekeeper use the casein and lacatalbumin it is necessary for them to watch out for lactose and over two- percent sodium. When the beekeepers use soy flour, they try to get the "debittered" soy flour that has been processed and retains some lipids, and toasted to knock out enzymes that interfere with the bees' digestion. Always make sure to check the data on the soy flour. The beekeeper will want to determine if the soy is a "high sucrose" variety or contains mostly stachyose. Stachyose is toxic to bees. Beekeepers will sometimes add a "feed yeast" like Torula to the pollen mixture to enhance the nutrients in the substitute. Most of them don't use it because of the high cost.

Pollen substitutes do not increase brood production as well as pollen sources brought in by the bees themselves. Because of the pollen substitute brood rearing will not stop all together should the weather stay bad for a while. A beekeeper will have a fatter bee when using a pollen substitute. There are some areas where pollen is scarce in the late summer and fall. If the beekeeper feeds the bees pollen substitute for a fatter bee, a fatter bee will winter better and rear more brood the next spring than their non-fed counterparts.

Alicia Moore

Bees are not fond of pollen substitutes. It must be place directly in contact with the bees and as close to the brood as possible. As long as the bees are bringing in a trickle of pollen the substitute will be eaten. If there is no pollen being brought in, the substitute will be ignored and will spoil over time. There are some commercially formulated pollen substitutes on the market that claim the pollen substitute is so attractive to the bees that they will eat it anytime the substitute is offered. No one has investigated those claims.

CHAPTER 6- KEEPING BEES IN A SUBURBAN AREA

If you want to keep bees in a populated area, you will need to know the basics of bee biology, property rights, and human psychology. It can be done with very few problems. Even in a city it is possible for bees to find enough pollen to feed them and produce a honey crop at harvest.

Beekeepers in the suburbs and cities need to manage their bees so they do not create a problem for the neighbors. Measures can be takes to alter the keep the bees from becoming a nuisance to other people. To do this we need to understand the circumstances, which cause bees to bother other people.

The bees flight pattern is one of the ways bees can be a problem for other people. When the bees leave their hives to gather food, they will fly 3-4 feet off the ground. You can prevent them from crossing paths of people walking in their flight path by planting a hedge or building a fence at least 6 feet tall. This forces the bees to fly above the fence. The hives can also be placed on the rooftop, which starts them out flying at a higher level than most people walk.

Fence, hedges, and rooftops also provide seclusion, which is very important. By keeping bees out of sight they will not be the target of vandalism or theft, also keeping bees out of sight will alleviate worried neighbors.

To keep the bees happy it is important for their hives have to be in a certain condition. A good location is for the hive to be in full sun all day, shaded bees will be more aggressive. The hives should be dry and the bottom boards angled so that water runs out of the hives. The hives need to be elevated with hive stands to keep the bees off the ground and to allow for airflow to keep the bottom board dry. Also with the hives 4 to 6 inches off the ground will make it less likely for grass and weeds to obstruct the view.

If you live in a congested area, a top entrance is probably not a good idea, especially during the summer. When ever a hive with a top entrance is opened and hive bodies moved, hundreds of confused bees will be fling around because their entrance is gone. This will probably worry you and your neighbors. By providing only a bottom entrance, and working from the side or from behind the hive, the bees are not impeded from flying home even when all the upper boxes are removed. Always keep the equipment in good repair. You don't want the cracks or chips in the hives providing extra holes for flight.

A bee only stings as a defense against intruders that might want to cause harm to the hive. Whenever a hive is open, the bees are in their most dangerous state.

During a nectar flow, many of the older workers will be in the field hunting for food. This is the best time to examine the colony. During the summer more bees will be in the hive and the situation can change, especially between the nectar flows. There can be some robbing going on at this time, which will make the bees even more defensive at any intrusion to their hive. Leaving the colony open for more than a few minutes can accelerate a robbing as can leaving cappings or honey exposed. It will become a necessity to reduce the entrance of a weak colony to prevent stronger hives attempt to rob from it. A honey flow will reduce the likelihood of robbing.

The mood of the bees can have a lot to do with the weather or the time of day. On the days of rainy weather, cool temperatures, early in the morning or late in the afternoon will be more likely to make them angry and they will attack. Always inspect them on warm, sunny days in the middle of the day when most of the bees are foraging.

Keep a constant warm water supply for the bees to cool the hive and dilute honey to feed t heir young. They will collect water from the closest water source. If you do not have a constant supply of shallow water for the bees, they will look for it somewhere else, like the neighbor's pool, birdbath or wading ponds. The bees are more likely to drown in those sources. If you have a water supply for them when they first fly out in spring, they will not go anywhere else for water. Once they find a water source, it is hard to keep them from going back to it.

A beekeeper must keep the bees in control every time the hive is open. A typical hive can house thousands of workers all capable of stinging. There are measures a beekeeper can take in the open that he cannot take in the city because of the closeness of other people.

Smoke is the most important tool for the beekeeper opening a hive. Smoke should be used in moderation, but the smoker should be capable of producing large volumes of smoke on short notice. The beekeeper must smoke the entrance of the hive, under the cover, and periodically smoke the frames while the hive is open. Try not to jar the hive or the frames as that may anger the bees, which will make it hard for a beekeeper to do his work. The beekeeper must work quickly and carefully. By going through the frames several times a year, the beekeeper keeps the frames movable. Remove any excess combs.

CHAPTER 7- ABOUT BACTERIAL DISEASES

There are two bacterial diseases that beekeepers must be on the lookout for they are American Foulbrood and European Foulbrood.

The American Foulbrood, also known as AFB, is the most serious of the bacterial diseases of honeybee brood and is caused by the bacterium Paenibacillus larvae. This disease is started and can be transferred only in the spore stage. The reason for the seriousness of the disease is the spores can remain alive and last for an undetermined length of time on beekeeper's equipment. It is highly contagious and spreads easily via contaminated equipment, hive tools, and beekeeper's hands. The best way to handle the American Foulbrood is to avoid it at all possibilities.

To detect the disease examine the larvae. Normal healthy larvae are white, but the infected broods turn chocolate-brown and melt into a gooey mass on the floor of the cell. The colonies will display a "pepper box symptom" as the disease progresses. The "pepper box symptom" is when the bees are capping the cells, the brood capping are perforated and sunken into the cell. When the larvae are brown and have not formed a hardened scale, the symptom of ropiness can be demonstrated. To do this, poke at stick into this mass, moisten it and withdraw it from the cell.

The contents will draw out like melted cheese, the ropiness, if AFB is present. As the dead larvae dries, it becomes a black scale that sticks tightly to the cell floor. These scales are difficult to remove and are site for re-infection. A single scale can contain one billion spores. It only takes 35 spores to trigger the disease. These scales are difficult to see and easily missed when purchasing used equipment. If you are around a colony that is extremely infected with American Foulbrood, it will emit a foul odor like a chicken coop. The colony dwindles and eventually collapses as more and more brood become infected and dies.

The beekeeper has an advantage if new equipment and tools can be purchased, install packaged bees and maintain them in total isolation from other apiaries, hive collections. Of course this is not realistic or practical, but it always makes good sense to practice sanitation, such as washing hands and hive tools regularly. Avoid using hive equipment of unknown history, and avoid feeding bees honey from an unknown source.

It is possible to breed bees that are genetically resistant to American Foulbrood and other diseases. One of the most important characteristics is the disease resistant bees is the ability to detect and remove from the colony abnormal cells of brood. The resistant queens are available from nationally advertised queen breeders. You will find the advertisements in the "American Bee Journal", "Bee Culture", and "Speedy Bee".

European Foulbrood, also known as EFB, is another of the bacterial diseases that affect the honeybee brood. There are some differences between the European Foulbrood and the American Foulbrood. The colonies infected with the American Foulbrood sometimes recover from the infection. The symptoms can sometimes be mistaken for those of the American Foulbrood, but there are some important differences. Instead of being a normal healthy white, the larvae with European Foulbrood are off-white, progressing into a brown, and are twisted in various positions in the cell. Larvae with European Foulbrood usually die before they are capped whereas with American Foulbrood die after they are capped.

The sanitation precautions recommended in the section on American Foulbrood also apply to the European Foulbrood. Bee stocks that are bred for resistance to diseases can be expected to minimize outbreaks of European Foulbrood. There are times at the onset of a strong nectar flow that the disease will go away on its own. The beekeeper may be able to control the disease by stimulating a nectar flow and by requeening the colony.

There is a preventative measure that can be used on either the American Foulbrood or the European Foulbrood, and is periodical treatments of the veterinary antibiotic TerramycinJ. It is fed as a mixture in either powdered sugar, sugar syrup, or in vegetable oil extender patties. It is very important to never feed the antibiotic within four weeks of a nectar flow to avoid contamination honey for human consumption.

The use of TerramycinJ in European Foulbrood infected colonies may actually be counterproductive because the medication permits those infected larvae to survive when they would have died. These survivors then are in the colony as a source of recontamination. If the infected larvae die instead, the house bees eject them from the hive and with them the source of the infection. The bacterium does not form long-surviving spores that will stay on the hive surfaces.

There has been recent evidence of the disease becoming resistant to the antibiotic. One of the suspected causes is the use of the oil extender patties as a method of medicating the bees. If the bees do not consume the patties rapidly, it leads to the antibiotic staying in the hive for weeks or even months. Until the use of the oil extender patties in the 1990's, resistance was not a problem. Beekeepers are now being told to remove uneaten patties after a month.

Sacbrood is a virus infection that is like a cold in humans. There is no known cure at this time. The best preventive measure is sanitation. Comb replacement and requeening the colony is the best response to the infection.

Beekeepers do not consider sacbrood a serious threat, however one larva killed by the sacbrood virus contains enough virus to kill over one million larvae. More research needs to be done on the sacbrood virus. It is unknown how the virus is transmitted to the larvae in nature, why severe outbreaks occur only during build-up season, or how the virus seems to return year after year.

Symptoms of sacbrood are partially uncapped cells scattered about the frame or capped cells that remain sealed after others have emerged. Diseased bees inside the cells will have darkened heads, which curl upward. The dead prepupa resembles a slipper inside the cell. Diseased prepupae fail to pupate and turn from pearl white to pale yellow to light brown and finally, dark brown. The skin is loose and flabby and the body watery. The dark brown bee becomes a wrinkled, brittle scale that is easily removed from the cells.

About Viruses and Fungal Diseases

Chronic Bee Paralysis is another of the viral infections that can plague a bee colony. Like all of the other bee viruses there is no cure or medication that can be taken to eliminate the infection, the only preventative measure is sanitation.

There are clearly defined symptoms with the Chronic Bee Paralysis. It only affects the adult bees. The symptoms are an abnormal trembling in the wings and body, the bee's inability to fly which forces them to crawl on the ground and crawl up the blade of grass in front of the hive. The abdomens will be bloated and the wings will be partially spread or seem dislocated. The infected bees will appear shiny and greasy because of the lack of hair, which has been confused with robbing bees. Also, the infected adult bees are

Alicia Moore

chewed on by the other bees and harassed by the guard bees at the entrance to the hive, which is also confused with signs of robbing. Adult bees will die within a few days of the onset of the disease. The virus is spread from bee to bee by prolonged bodily contact or rubbing which causes many hairs to break exposing live tissue. The virus cannot be transmitted by food exchange of the bees. It takes many millions of virus particles are required to cause paralysis when given to a bee in food. Requeening is a good practice if symptoms appear.

Another virus that bees are susceptible to is the Black queen cell virus. It is associated with Nosema disease and causes the death of queen larvae or prepupae after their cells are sealed. Th larva will then turn black along with the walls of the cell. Treating colonies with Fumidil-B to control Nosema may help keep prevent this disease.

A fungal disease that plagues the bee colonies is called Chalkbrood. The fungus that causes Chalkbrood is called Ascosphaera apis; it was discovered here in the United States in 1968. The fungus spores must be ingested in order for infection to occur. It only infects larvae 3 or 4 days old. There are no chemical treatments for this disease. However, bee breeding and good management can control it. The infected larvae are quickly covered with the white cotton-like mycelium of the fungus, which eventually fills the entire cell. The white/gray mass soon will harden into a hard, shrunken mummy, which is easily removed from the cell. The larvae in the cell will look like a piece of chalk.

The bee bred to be resistant to this disease can help minimize outbreaks of t his disease. Another way to cut down on the number of outbreaks of the disease is to maintain a warm, dry hive interior. If the hives are drafty, damp, lying in low spots or in heavily overgrown area, they are more susceptible to chalkbrood disease. Rain water need to run out of the hive instead of

35

accumulating, so stand the hive with it leaning forward slightly. If a hive gets moist, prop the lid of the hive open to air out the interior. Old equipment should be replaced or repaired if it has large holes that permit entry of moisture and drafts.

There is a possibility of genetic susceptibility or old combs that are harboring spores of the disease if the colonies have recurring problem with the disease that are not easily traced to season or management practices. Old combs should be replaced periodically to improve brood production.

About Varroa Mites

Varroa mites were first discovered in the United States in 1987, and then the mites were detected in North Carolina three years later. The mites have since spread throughout the rest of the country. They are considered to be the most serious pest of honeybees worldwide. Infested colonies will die within 1 to 2 years unless the beekeeper takes the necessary actions to rid the colony of the mites.

The Varroa mites are external parasites of the drone and worker bees. They prefer drones, but will infect the workers also. Varroa mites are visible with the naked eye and look somewhat like a tick. The mated female moves into a brood cell with older bee larvae. Mites will feed on the larvae food or puncture the larval body and feed on the bee's blood. The mated female mite will lay an egg every 36 hours on the side of the cell. The first egg will be unfertilized and develop into a male. The other eggs are fertilized will hatch into females. The young mites feed on the developing pupa. The young females will then mate with the male and emerge from the cell when the bee emerges. The female mites will then enter another cell or attach themselves to an adult bee to feed on. The Varroa mites are transported from colony to colony by drifting or robbing bees.

Alicia Moore

There are visible symptoms of the damage from the mites on the newly emerged bees, which is due to the mites feeding on the immature bee in the cell. The newly emerged bee will be smaller than normal, have crumpled or disjointed wings, and shortened abdomens. The life span of the infected bee is also shortened. Severe infestations from the mite within the cell, which is several mated adult female mites in one cell, can cause death to the pupa. Other symptoms of mite infestations are the rapid decline of the colony, reduced adult bee population, evacuation of the hive by crawling bees, queen's lack of performance, spotty brood, and abnormal brood.

Detection is the first step to control. There are methods used to detect the presence of the Varroa mites as follows:

Extract drone brood when present and visually examine larvae and cells for mites. There are visible against a light colored background.

Fill a quart jar about 1/4 full of live bees. Cover and insert a 2-second blast from an aerosol ether-based engine starter fluid or aerosol oil cooking spray. Shake the jar for 20 seconds. Turn the jar on its side and rotate slowly and look for mites clinging to the sides of the jar. If you do not spot any mites, remove the bees and rinse in alcohol. Shake and remove the bees so you can examine the alcohol.

The best and most reliable method is to use Apistan@ (fluvalinate) strips or US: Check Mite+ strips.. Place a piece of waxed or white paper sprayed with aerosol oil cooking spray and covered with 8-8 squares/inch of mesh wire on the bottom board. Insert strips according to label directions. Check the paper in one hour. If there are no mites, check again the next day.

You can request a free inspection from you local NCDA bee inspector.

Never treat during a nectar flow because the chemicals can contaminate the honey and never leave strips in hives after the recommended time this can cause sublethal doses of the chemical. However, if mites are detected, you may need to treat to save your colony.

In recent years mite have become resistant to Apistan strips and has become a problem throughout the world. Therefore, rotating chemical, delaying treatment and using cultural control are recommended to manage mites in a more bearable fashion.

Delaying treatment can be accomplished if you monitor the level of Varroa mite infestation in your colonies. There are ways to check the colony for the number of mites present. Knowing the level of infestation in your colonies will help you determining whether treat is required immediately or if it can wait until after the nectar flow season has passed.

About Tracheal Mites

First detected in the United States in 1984 the Tracheal mite has caused the loss of tens of thousands of colonies and millions of dollars. The tracheal mite will infest the tracheal system of the adult honey bee, they prefer adult bees less than four days old. Levels seem to be at the highest during the winter and spring. Once they are on the bee, the mites are attracted to the carbon dioxide exhaled and enter the spiracles located on the thorax, which lead to the tracheal system. They will puncture the wall of the trachea and suck the blood of the bee. Once in the tracheal system the mites live, breed and la eggs. The adult and the eggs plug the tubes of the trachea, which impairs oxygen intact of the bee. Since they puncture the trachea in order to feed, they will

Alicia Moore

spread secondary diseases and pathogens. The bee dies from the disruption to respiration damage to the trachea, and from the loss of blood. Once over 30 percent of the population are infected with tracheal mites, honey production may be reduced. The likelihood of winter survival decreases with increasing infestation of the mite. Mites are transmitted from bee to bee within a colony by robbing or drifting bees.

Infested bees will be seen leaving the colony and crawling on the grass just outside the hive. They will crawl up the blades of grass or the hive, fall back down and try again. The wings will be disjointed and the bees will be unable to fly. If you are unsure about a tracheal mite infestation, send sample bees in alcohol to your local county extension agent for verification.

One method of preventing tracheal mites is an oil extender patty. It consists of two parts sugar to one part vegetable shortening. Make a small patty about four inches in diameter. Sandwich it between was paper. Cut the wax paper around the edges so the bees have access to the patty. Place the patty on top of the frames in the center within the hive body. The bees will be attracted to the sugar and get oil on their body. The oil makes it difficult for the mites to identify suitable bee hosts. The oil patties will not contaminate the honey supply so they can be used for prolonged periods.

There is one other method for controlling tracheal mite infestations. Menthol can be used and is available in most bee supply stores. The temperature must be above 60⬚F in order for the menthol to work. The bees breathe the vapor, which dehydrates the mites. Menthol must be removed during a nectar flow so that the honey is not contaminated.

The Small Hive Beetle

You will find the adult and larvae of the small hive beetle are found in active beehives and stored bee equipment where they feed on pollen and honey. The small hive beetle is native to Africa where it requires 38-81 days to develop from egg to adult. Beetle larvae on not spin webs or cocoons in the beehive but rather pupate in the ground outside the hive. This first record of this beetle in the Western Hemisphere was determined from a commercial apiary in Florida in May 1998.

The small hive beetle behaved as a scavenger of weakened colonies in Africa. They were relegated to secondary pest status. Here in Florida it has not been the case. The apiaries suffered extensive damage and colony loss. Beetle larvae tunneled through combs, killing bee brood and ruining combs. Bees in Florid have abandoned combs and entire colonies once they are infested. The beetles would defecate in the honey causing it to ferment, producing a frothy mess in supers and honey houses. Honey contaminated can no longer be sold and cannot be used as bee feed. In heavily infested apiaries in Florida, larvae could be seen crawling out of the colony entrances or across honey house floors by the thousands trying to reach soil to dig in and complete their development. It has been cause for some concern regarding the beetles behavior in Florida compared to its behavior in Africa.

The following precautions are suggested to help maintain control of the beetle.

1. Make sure the area around the honey house is clean. Extract honey from filled supers as soon as possible rather than let them stand too long. Leaving the cappings exposed for too long is another bad idea. Beetles can multiple rapidly in stored honey, because the honey is away from the protective bees.

2. Avoid stacking infested supers in strong colonies.

Alicia Moore

3.   Notice when *supering colonies are making splits, exchanging combs or use of *Porter bee escapes can spread the beetles or provide room for beetles to become established away from the cluster of protective bees.

4.   Watch colonies for sanitary behavior, such as bees showing the ability of ridding themselves of the larvae and adult small hive beetle. Breed queen lines found to be beetle resistant.

5.   See if it is possible to trap the beetle larvae as they make the trek to reach the soil. Moving colonies might be useful in keeping a beetle population from growing. The beetle may be adverse to certain soils. In this case fire ants may be a predator for the beetle larvae as they are pupating.

6.   Bees will not normally clean-up equipment or supers full of beetle-fermented honey. Bees, however, will finish the job after the beekeeper fist washes out as much honey as possible with a high-pressure hose.

7.   By treating the soil in front of the affected hive with a soil insecticide the larvae may not reach adult stage.

8.   Treat colonies with Check Mite+ beehive pest control strip according to label instructions.

*supering - the filling of the supers with excess honey

*Porter bee escape - originally designed to clear bees from supers that were to be extracted.

About Nosema

Nosema is the most widespread of the adult honey bee diseases. A single celled animal named Nosema apis, a small, unicellular parasite specific to the honeybee, causes it. Nosema cannot exist in a laboratory culture, as with most bacteria and fungi. It will only thrive and multiply in the epithelial cells of the honey bee ventriculus which causes dysentery. Queens, drones and workers are all susceptible to Nosema. The spores of the Nosema must be ingested for the bee to be infected. The spore takes root in the midgut, where they will penetrate a midgut cell and grow by absorbing nutrients from that cell.

The parasite will increase in size until it is large enough to divide in half. Each new parasite will continue to feed on the nutrients of the cell until they are depleted. In a matter of time, about 6 to 10 days, 100 new spores are formed in the infected cell. The infected cell when depleted of all the nutrients ruptures releasing all the newly formed spores into the midgut to start the process again. The damaged intestinal tissue is susceptible to secondary diseases. Dysentery is a common symptom of this disease. You will be able to spot the dysentery on the outside of the hive by the little brown spots, but the diseased bees will also defecate inside the hive. contaminating combs with millions of infectious spores. The disease is spread to other colony members through fecal matter.

Nosema having infected one bee will be spread to others in the colony. The disease lowers the life span of the bees. If you have a colony of bees infected with Nosema in late fall, come spring it is likely that most of the colony will have died off.

Nosema is a difficult disease to diganose without using laboratory equipment. Decapitating a bee and pulling out the last abdominal segments usually will remove the intestinal tract while still intact. An infected midgut will become swollen, whitish and lose its visible constrictions. However, other causes of dysentery, such as ingesting honeydew, fermented syrups, etc. can result in similar intestinal changes.

Treatment for Nosema is based on the most appropriate times to prevent comb contamination and to prevent the development of disease in bees that clean up fecal deposits from combs while they are still trying to expand the brood nest. A few bees are always infected, but the diseased late season bees are the only one of any concern. If they develop high levels of infection, they defecate on the combs in October, November, and December, and then they die. The use of fumagillin has been field tested by some beekeepers with acceptable results. When treating use the manufacturer's instructions.

CHAPTER 8- ABOUT THE DISAPPEARING BEES

News agencies started reporting on a disturbing phenomenon in the bee population, in the spring of 2007. It was reported beekeepers were visiting their hives to discover that their bees had disappeared. The queen and a few newly hatched bees were all that remained. The presence of predators feeding on the bees did not leave any evidence of having been there. There was no evidence of dead bees from bee diseases either. Based on the lack of evidence, it seemed unlikely that the bees had gotten sick and died. However, many beekeepers reported that moths, animals, and other bees steered clear of the newly emptied nests. This is a normal reaction when bees die from disease or chemical contamination.

The news reports were alarming. They described beekeepers losing more than half of their bees and explained the importance of honeybees in the pollination of food crops. Some of the articles implied with the disappearance of the bees widespread starvation would follow. The disappearing of bees or otherwise called "Colony Collapse Disorder: is a real phenomenon. It has the potential to impact food and honey production, but it is more complex than it has been reported.

The colony collapse disorder has had an effect primarily on the domestic, commercial honeybees. These bees are raised exclusively for producing honey and pollinating crops. It also seems to effect bees from hives that are moved from place to place to pollinate crops. Of the overall bee population, the commercial honeybees make up only a small portion. Africanized honeybees, along with other types of bees, do not seem to be affected.

Alicia Moore

Also, this is not the first time the honeybee population has suddenly and unexpectedly declined. In the last 100 years beekeepers have reported sharp decreases in their hive populations several time. In 1915, beekeepers in several states reported substantial bee losses. The condition became known as the "Disappearing Disease". It was not named for the bees disappearing, but because the condition was limited and did not happen again.

Researchers never determined the cause for Disappearing Disease or the declines in bee population, and the causes are still unclear today for the colony collapse disorder. Several possibilities have been ruled out because they are not present in all of the affected colonies. The bees in the affected colonies were all feed using different methods, mites and other pests were controlled in a different way. The bees did not even come from the same supplier. The work group investigating the phenomenon does not suspect genetically altered crops to be the problem.

There are some theories on the causes of colony collapse disorder.

The process of transporting bees over long distances in order to pollinate crops may cause stress, which has depressed the bees' immune system, exposed them to additional diseases or affected their navigational abilities.

Mites generally feeding on the bees may be exposing the bees to an unknown virus. Mites have caused colony collapse in the past, but they have also left evidence, which is not the case in colony collapse disorders.

One common theory regarding cell phones as the culprit, but it has been discounted. This theory made the news in April, 2007, "The Independent" who featured the article about a study being done on the cell phones and linking them to the bee disappearance, they failed to dig deep enough for their story. The study was not related to cell phones, but was on the electromagnetic energy coming from the base units of cordless phones. A cordless phone uses a different wavelength than the cell phone.

It is unknown exactly where the honeybee species is headed or exactly how the drop in the population of the bee will affect the world's food supply. The drop in population in all likelihood not lead to the sudden extinction of the human race, it is going to have an l effect on what we eat if it continues.

Bee Stings

As a beekeeper you will be subjected to bee stings. They will decrease in time, as you become more adept at the handling of bees. If you should be stung, you will need to know what to do. When a bee stings you the stinger will remain behind because of the barbs on the stinger. DO NOT pull the stinger out this only release more of the bee venom into the sting site. Scrap the stinger out. Use a fingernail or even the hive tool to remove the stinger.

The stinger contains glands that secrete chemicals that is an alarm odor. Because of this, if you are still around the hives, other bees will either sting the same area or buzz around it. Puff some smoke on the sting area and remove yourself away from the hives. Wash the site with water to remove the chemical causing the odor. Washing isn't usually necessary because by scraping the stinger away and removing it the alarm chemicals go with it.

You may want to use a sting relief medication on the site, as it will hurt for a while. Otherwise a cool compress will provide some relief. There are some home remedies you can use that will help alleviate the discomfort.

Alicia Moore

You can apply a solution of 1 part meat tenderizer to 4 parts water. Papain is the enzyme in meat tenderizer that will break down the protein of the bee venom, which causes the pain and the itching. Leave this on for no more than 30 minutes.

You can also try antiperspirant; the aluminum chlorohydrate reduces the effects of the bee venom, but is not as effect.

Applying cold by using ice or cool water for 10 to 30 minutes after the sting blunts the body's allergic response.

Placing a raw onion on the sting will draw the poison from the wound, helping you get relief easily

Benadryl or any other antihistamine taken by mouth can give some added relief, and help prevent the reaction from spreading.

Calamine lotion or hydrocortisone creams can have a similar effect. As will as making a paste made of baking soda and water, leave on for 10 to 20 minutes.

Pain relievers such as Advil or Tylenol can be administered for pain relief.

These are just some of the home remedies.

Pain and swelling are common reactions to a bee sting. You are not having an allergic reaction. After a day or so the sting will itch. Don't scratch because it will become worse and could get infected. The swelling and itching may persist for a day or two following the bee sting. You should be over the effect of the sting in about 4 to 5 days.

If you are having an allergic reaction you will experience difficulty in breathing and swallowing, dizziness, a rapid heartbeat, nausea, cramps and vomiting, shock and headaches. Seek medical attention immediately.

If you receive multiple stings, it may be a sign of aggressive bees. Use your smoke and close the hive as quickly as possible without causing the bees any more alarm. If there is a specific reason for the aggressive behavior of the bees, it may be eliminated. Allow the bees the opportunity to calm down and they may become more manageable. Multiple stings only create more discomfort. They are not more severe to anyone even an allergic person, with the allergic person several stings is just as bad as one sting..

CHAPTER 9- THE PROCESSING OF HONEY

If the world were perfect, supers would be removed and taken to the honey house, to start the processing. Here is this real world the honey can be left in the super too long. Then you have several dangers to consider. Honey remaining in the super can be subject to robbing, by insects or mice, damage by wax moth, and fermentation.

Supers can be stacked in a garage, an outdoor workshop or a room indoors, provided it is clean, dry and protected from excessive heat. Stored honey can be tainted by the odors from paint, chemicals and even cooking.

The stored supers with honey are still at risk of dangers from ants, earwigs, bees and wasps. Plus physical and chemical changes can take place in honey that has been in storage for a prolonged length of time.

The main factor in honey is the water content. Honey with more than 21% water content with the exception of heather or clover honey is not fit for sale, except for industrial use. Honey when exposed to the air will attract moisture from the atmosphere and in very dry, warm atmosphere, the honey will lose water, and the quality will improve. Sign to watch for are watery honey running from open cells, bubbly honey, and honey weeping through cappings. One or more cells in this condition in a super will not ruin the lot. You have not wasted your time extracting it for human consumption. However, the bees will readily take it back as a feed, with no ill effects.

A honey room for the purpose of processing honey has some requirements. First thing is hygiene; Floors and surfaces need to be washable. A toilet facility needs to be available along with washing facilities. Hot and cold water may not be imperative, but are strongly recommended. When family and friends extract honey only for consumption and not sold on the market, the odd bee wing or lump of wax is not a disaster.

However, when it comes to honey for sale, if unsatisfactory in any way, can bring a visit from a Trading Standards officer to scrutinize every part of the operation. If keeping bees and wasps out is a difficult task, to may be worth doing this process at night when the foragers are not flying. After working during the night, all the honey can be packed away, supers sealed and equipment washed before enough bees discover the feast.

The thickness of liquid honey changes with temperature- the higher the temperature, the runnier the honey. The lower the temperature the thicker the honey making it difficult or even impossible to remove from the extractor. As a rule of thumb the temperature should range between 70F and 95F. The frames will empty quickly and setting or "ripening" is more, thorough. Air escapes easily with less froth, and heavier particles drop quickly. The honey room layout should be planned so that there is an easy flow from one task to the next. Lifting and moving of supers and frames should be minimized.

Honey and wax will inevitable reach every corner of the room, floor, door handles, taps-anything touched by foot or hand will be sticky. Throughout the processing, keep handy one bucket of warm soapy water for washing surfaces. This will help keep the mess under control, and another container for washing hands and utensils. Wax is removable with a sharp stick when the room is cooler.

Alicia Moore

As a beekeeper just starting out it can be extremely confusing with all the hives, frames and even bees, and that doesn't even include the honey extracting equipment. For a beekeeper with only one hive it may not cost effective to lay out the money for elaborate equipment. It is perfectly practical to enjoy the honey crop using basic kitchen tools. Before a super is put on the hive in the spring, the decision has to be made how to harvest the honey. The options are:

  a) Cut comb honey.
  b) Section honey.
  c) Extracted honey.

Cut comb honey is cut out of the frame and packed in 8 oz. and 12 oz. pieces. It is eaten with the wax comb, and is one of the best ways to present honey as aromas and flavors are unimpaired by extracting and heating. Granulated honey in comb is not very attractive to most customers.

To the beginner who does not have access to an extractor, this method is attractive, because a very small amount of equipment is required. To cut comb honey the super frames should be fitted with "thin super " or "extra thin" foundation. A whole sheet is usually used for each frame. A 25 to 50 mm deep full-width starter strip may be used instead. Cut comb containers commonly used can comfortably hold a comb about 40 mm thick.

Examine the frame before cutting to decide which side of the comb has the better appearance. Lay the frame on a clean tray, and the whole comb cut out of the frame with a sharp knife. Only the best parts of the comb can be used. The hollow parts at the edge should not be used and uncapped cells kept to a minimum. A sharp kitchen knife, a cheese wire, or a stainless steel comb cutter can be used to cut the combs. All portions of cut comb should stand on a grid to let the honey drain from the outside cut cells. A piece of comb honey swimming in its container in liquid honey is poor presentation. Because heather honey is a gel it can be packaged straight away. The best storage for comb honey is in a deep freeze, in special plastic boxes, where comb will keep indefinitely. Freezing packaged comb honey will also kill any wax moth eggs and larvae. Comb honey stored in any other fashion must be examined regularly for signs of deterioration. Another development of comb honey is chunk honey. Chunk honey is a piece of cut comb is put in a jar and surrounded with a clear runny honey, producing what is an attractive presentation.

Wax cappings are a valuable by product of extracting. After cappings have dripped dry, wash them in water to remove all honey. Melt the cappings, strain the wax through nylon and pour it into bread pans or a similar mold. Supply companies can render you beeswax bricks into new foundation at considerable savings.

An experience bee craftsman accomplishes section honey. Section honey is the finest and traditional way of presenting honey. There are tricks and quirks to this method that demand great attention. If you are interested in learning the craftsmanship of this type of honey presentation, you will have to get specialized books or literature on the subject. It is so detailed it can not be covered and given the justice it deserves in a small publication.

It is possible to extract honey without the assistance of a centrifugal extractor, by just using basic kitchen implements to cope with one or more supers. It will be time consuming, sticky and inefficient, but if it means that the beekeeper's family can obtain some benefit from his or her obsession, it will be worthwhile.

This method of extraction requires that the comb, cappings, cells, and honey to be scraped from the frame. A large table spoon or serving spoon handled carefully will allow the foundation to be left intact, while both sides are scraped reasonable dry. A few holes here and there will not matter to the bees who will patch it up later. The honey and wax should be mashed up in a clean basin or bucket, then tipped into a sieve or similar strainer and left to drain for at least overnight, but possible even for days. The wax left in the strainer will still contain a lot of honey, which is best fed back to the bees, by diluting with warm water, and putting the mix, wax and liquid, into any kind of feeder.

The warmer the honey the easier it runs. So prior to the extracting it is best to warm the honey. A pile of supers with a large amount of honey will not warm up enough by simply bringing them into a warm room for an hour or so. It might take as many as two days to do the job. The moisture content of the honey will be reduced during a warming process. To accomplish the warming of the honey, it is possible to pile the supers in staggered stacks with a fan heater directed towards them. There are some drawbacks to keep in mind. They are:

a) Heating will remove some of the compounds that give the honey its unique flavor and aroma. Prolonged heat can darken and damage the honey. There are tests to be used to distinguish overheated honey.

b) The wax will soften making uncapping more difficult, with cell walls dragged along by the knife. This will happen at 400°C, at 450°C combs will soften and collapse, and at 630°C wax will melt.

Each frame is lifted from the super with one lug located on a bar over a bucket or tray or tank. The capping is then removed by using a cold knife, cappings scratcher, cranked uncapping fork, or electric knife. The amount of honey mixed with the wax cappings will vary, depending on the method used for the uncappings.

a. The simplest way, is by uncapping into a bucket, basin or uncapping tray and then by gravity straining with a strainer or sieve. A filter bag, tailored to a 70 lb. plastic tank is typically used. The honey left in the wax cappings can be washed out and used for making mead (a honey wine) or fed back to the bees.

b. Using a heated tray while uncapping, the wax and honey can be separated and processed at the same time will cut out a lot of the sticky work. The stainless steel tray has an electrically heated water jacket. Honey will run down the surface, while the wax is held back and gradually melts. The honey and the wax will end up in the same bucket. The wax solidifying and floating on top of the honey will separate the wax from the honey.

There are other processes for separating honey and wax that require elaborate equipment

Equipment used for Honey Processing

Centrifugal extractor is based on the same principal of a centrifuge. The frame is rotated in order to throw out the honey of the super. As a beginner you may be able to borrow one or rent one from the local association. If you are planning on making a purchase of one, you will have some choices to make. You can choice a tangential or radial, plastic or stainless steel, and manual or electric.

Let's look at tangential first. In a tangential machine the frames lie almost against the barrel of the drum. The outer side of the frame is part that empties when spinning. The machine is evenly loaded. Then it spins until about half the outer side has been extracted. You will be able to see tiny dots of honey flying from the frame and hitting the barrel. Turn the frames around so that the other side of the frame is facing outward. The spin the machine again until all the honey has spun out. The frame is turned one last time and spun for the final removal of the honey. This method prevents the combs breaking from the middle being full and the outer side empty. Each frame does have to be handled four times and the machine stopped and started 3 times.

The handling time using this machine is a disadvantage; however, the extraction of the honey is more thorough than other machines. It is the most compact extractor available, so therefore cheaper than other machine. If you are extracting heather honey, this is the only type of machine to cope with it.

The frames sit between rings, arranged like the spokes of a wheel in a radial machine. The extraction takes place on both sides at the same time, so there is no need to move the frames once they have been loaded. The radial machine is larger than the tangential machine. This is to ensure that the frames are far enough from the center to extract evenly. Because of the size of the machine it is capable of handling a lot more frames than a tangential. In both machines there is not major difference in rotation direction, but the electric radial machines have a reverse position to remove a little more honey from the cells and dry out the combs.

The traditional material used in the construction of the machines is usually tin-plated steel. A good quality tin-plated steel will last for many years unless it starts rusting. Once the machine starts rusting there is very little to be done about the rust. The barrel can no longer be used for the processing of a food product. The tin-plated extractors have been replaced with plastic and stainless steel barrels. If you get a choice, stainless steel is more durable than plastic.

If you are only extracting honey from two or three hives, a manual extractor will do the job. If you have a considerable amount of hives, the manual machine can become extremely tiring to use. When it comes to making a choice, it may depend on the money available, the stamina and the outlook of the beekeeper. The electric extractor will not only save you labor, but also reduces the time taken. The beekeeper could be uncapping while the extractor is running with the previous load.

Alicia Moore

ABOUT THE AUTHOR

Two critical secrets that Alicia Moore discovered in beekeeping are bees, like most creatures, need water to survive. Consider installing a fountain or a birdbath with small fountain. Many birdbaths come with solar panels that keep the water moving. Stagnant water often draws mosquitoes. Place a few small stones in the bird bath or fountain for bees as well as butterflies. Next, instead of using chemical pesticides that may kill off bees, try alternatives to pesticides. Certain plants naturally repel various types of insects or pests. Plant garlic to repel Japanese beetles, weevil, aphids and spider mites. Marigolds repel cucumber beetles and basil repels tomato horn worms. Alicia reveals more secrets in her book. Grab a copy now!